IMAGES
of Aviation

HILL AIR
FORCE BASE

IMAGES
of Aviation

HILL AIR
FORCE BASE

Lt. Col. George A. Larson, USAF (Ret.)

ARCADIA
PUBLISHING

Copyright © 2021 by Lt. Col. George A. Larson, USAF (Ret.)
ISBN 978-1-4671-0643-6

Published by Arcadia Publishing
Charleston, South Carolina

Printed in the United States of America

Library of Congress Control Number: 2020946682

For all general information, please contact Arcadia Publishing:
Telephone 843-853-2070
Fax 843-853-0044
E-mail sales@arcadiapublishing.com
For customer service and orders:
Toll-Free 1-888-313-2665

Visit us on the Internet at www.arcadiapublishing.com

To the men and women, both past and present, who have
served at Hill Air Force Base—thank you for your service.

CONTENTS

ACKNOWLEDGMENTS

I want to thank the US Air Force Book Program in New York City, and Lt. Col. Jamie L. Humphries, director, US Air Force National Media Engagement. I want to recognize Derick Kaufman and Air Force Material Command Public Affairs for support. Jonathan F. Bingham, 75th Air Base Wing historian, provided historical and current photographs, information, and suggestions on what and how to cover the history of Hill Air Force Base. His coordination was essential in organizing the book into a comprehensive and wide-ranging history from the 1920s to the present. Without his efforts, this book would not have been possible. The Wendover Historic Airfield Association provided information on the desert bombing range and its role in ending World War II through the 509th Composite Group dropping two atomic bombs on Japan.

INTRODUCTION

The history of Hill Air Force Base and the Air Logistics Center begins in 1920 with the Ogden Ordnance Reserve Depot, designated to store surplus World War I ordnance on 1,410 acres of land purchased by the federal government. Construction was completed in 1922. In 1927, the depot was redesignated the Ogden Ordnance Depot and used as a general storage depot until 1936 under reduced military funding because of the Great Depression.

In February 1934, Pres. Franklin D. Roosevelt tasked the US Army Air Corps to deliver airmail, flying 12 routes until June 1934. In July of that year, the Air Corps recommended that a depot be constructed in the Rocky Mountain region. Congress passed the Wilcox Act in 1935, authorizing site selection and construction of seven permanent Air Corps stations, one to be the Rocky Mountain Air Depot. In 1936, rehabilitation and reconstruction began on the Ogden Arsenal for the manufacture and storage of bombs and shells.

By April 1939, the federal government had acquired 3,000 acres from the Ogden Chamber of Commerce to construct the Ogden Air Depot. The Military Appropriation Bill of 1940 passed in 1939 with funding for the Rocky Mountain Depot. On December 1, 1939, the depot was named to honor US Army Air Corps major Pete Hill, who died at Wright Field in Dayton, Ohio, while piloting the first Air Corps XB-17. Though preliminary grading had started in November 1938, paving of the four 7,500-foot-long, 150-foot-wide runways did not begin until July 1940. This was done at an immense cost of $1,358,093, with an additional expense of $62,750 in 1943 for taxiway grading and paving. In January 1941, hiring of the initial civilian employees began; during World War II, base personnel grew to 15,780 civilians and 6,000 military.

During World War II, Hill Field and the air depot became an important Air Corps facility, which on June 20, 1941, became an Army Air Forces maintenance and supply base. It supported many different aircraft for structural repair, engine overhaul, and spare parts supply.

On July 15, 1943, Ogden's first contingent of the Women's Army Auxiliary Corp (WAAC) arrived on the base, designated the 907th WAAC Post Headquarters Company. On October 1, 1943, the unit was inactivated and its personnel became part of the WAC Detachment Number 1, Ogden Air Service Command, 482nd Base Headquarters and Air Base Squadron.

In 1944, Hill Field became a storage depot for surplus World War II aircraft and equipment. During the war, Hill Field ordnance personnel supported the Wendover Army Air base bombing range 80 miles west of Salt Lake City. By 1947, more than $200 million in aircraft had been processed and preserved for possible future use, with surplus aircraft taken apart for spare parts. On September 27, 1947, the US Air Force was created as an independent branch of the US military. Hill Field became Hill Air Force Base on February 5, 1948.

After the invasion of South Korea by the North Korean Peoples Army on June 25, 1950, Hill Air Force Base supported logistics efforts for the Korean War. In 1953, the Ogden Air Material Area had maintenance responsibility for the first generation of Air Force strategic missiles. On April 1, 1955, Hill's size doubled after the Department of Defense transferred the Ogden Arsenal from the Army to the Air Force, now referred to as the West Area, with its 600 buildings and specialized structures. In 1958, the Ogden Air Material Area assumed support for the Convair SM-65 Atlas intercontinental ballistic missile (ICBM), and Boeing SM-80 Minuteman ICBM. Hill Air Force Base became the single assembly and recycling site for the Minuteman. In the 1960s, Hill's support extended to Boeing's on-base Minuteman facilities.

Hill Air Force Base supported the McDonnell F-4 Phantom II, Martin Marietta Aerospace Titan ICBM, and Hughes Aircraft ASGN-65 air-to-ground missile. Hill was involved in support for the Vietnam War, delivering hundreds of tons of military supplies to South Vietnam by Douglas C-124 Globemaster II, Lockheed C-130 Hercules, Lockheed C-141 Starlifter, and Douglas C-133 Cargomaster transport aircraft. The base also assumed responsibilities for the Convair B-58 Hustler delta-wing supersonic nuclear bomber and the General Dynamics F-111's landing gear components.

The Ogden Air Logistics Center became the system manager for the General Dynamics F-16 Fighting Falcon and proposed the Advanced Intercontinental Ballistic Missile (the Minuteman's proposed replacement, cancelled in 1967), and Fairchild Industries A-10 Thunderbolt II ground attack aircraft in 1970. The center also managed the McDonnell Douglas F-15 Eagle and field testing of the UH-1H Iroquois helicopter.

In 1972, production of the first variant of the supersonic air-to-ground Short Range Attack Missile was delivered from the Boeing Air Force Plant 77 on Hill Air Force Base. The base had logistics responsibilities for Alaska, western Canada, Idaho, Montana, North and South Dakota, Wyoming, Utah, Colorado, Arizona, and New Mexico.

The 1980s brought responsibilities for the BGM-109G Ground Launched Cruise Missile. The base was assigned repair projects on the North American Bronco and Lockheed C-130 Hercules aircraft. In 1982, the Ogden Air Logistics Center was designated logistics system manager for the Martin Marietta Peacekeeper ICBM, designed as a replacement for the Minuteman but cancelled and pulled from alert at F.E. Warren Air Force Base. From 1990 to 1991, the Ogden Air Logistics Center and Hill Air Force Base's tenant units supported Operations Desert Shield and Desert Storm, and new air support missions. The 388th and 419th Fighter Wings provided combat aerial support on the Lockheed Martin F-16 Fighting Falcons.

After the terrorist attacks on September 11, 2001, the Ogden Air Logistics Center began a massive support effort for the war on terrorism through engineering, sustainment, logistics management, and maintenance support for the Minuteman III ICBM and Air Force aircraft. The 388th and 419th Fighter Wings deployed to Southwest Asia and the Middle East from 2000 to 2010.

In July 2012, the Air Force Material Command, Ogden Air Logistics Center's parent command, reorganized and was reduced from 12 to five logistic centers. On October 1, 2012, Air Force Material Command assigned the complex to Air Force System Command, and the majority of the acquisition functions and personnel once present in the Aerospace Sustainment Directorate and assigned to the Ogden Air Logistics Center now worked for the Air Force Logistics Center Material Command. The 309th Maintenance Wing was inactivated and its functions reassigned. The 75th Air Base Wing was assigned to Air Force System Command.

In 2013, the Air Force consolidated F-22 Raptor depot maintenance to the Ogden Air Logistics Center and base operations for the Air Force's first Lockheed Martin F-35A Lighting II fifth-generation fighter. In 2014, Hill Air Force Base started F-15 depot maintenance.

On August 27, 2019, Northrop Grumman conducted the ground breaking for its new facility, the Northrop Grumman Ray Innovation Center near Hill Air Force Base to serve as its workforce headquarters and nationwide team support for its Ground Based Strategic Deterrent Program, the replacement for the Minuteman III.

The 388th Fighter Wing, 4th Fighter Squadron received its first F-35A on September 27, 2017. On December 17, 2018, the wing received its last of 78 F-35As, equipping the 4th, 34th, and 421st Fighter Squadrons, each operating 24 primary and six backup F-35As. On November 5, 2019, an F-35A was the first aircraft to take off from Hill Air Force Base's renovated runway after a nine-month, $44.6 million construction project. Repairs were made to the 13,500-foot runway, which remained operational throughout the construction project, including wider shoulders and taxiway at the south end, new airfield signs, new electronic wiring that included runway lighting, new runway surfacing, and concrete pavement repairs.

Hill Air Force Base is under Air Force Material Command, with 75th Air Base Wing host units that consist of Air Force Sustainment Center with the Ogden Air Logistics complex, 448th Supply Management Life Management Center, and Air Force Nuclear Weapons Center. The 75th Air Base Wing oversees one million acres and more than 1,700 facilities valued at $4 billion while providing installation support for the Ogden Air Logistics Complex, Air Force Life Cycle Management Center, Air Force Nuclear Weapons Center, Air Force active duty 388th and reserve 419th Fighter Wings, and more than 50 mission partners. Air Combat Command controls the 388th Fighter Wing (active wing) and reserve 419th Fighter Wing. The 75th Air Base Wing has support responsibility for the operation of the Utah Test and Training Range. Located in Utah's west desert, the airspace is situated on 2.3 million acres of land and contains the largest block of overland contiguous special-use airspace in the continental United States.

As of 2018, the Hill Air Force Base workforce consisted of 25,709 personnel (5,788 military, 3,621 military dependents, and 16,300 civilians). The base has an annual payroll of $1.43 billion, with annual expenditures of $760 million. Hill Air Force Base creates $1.38 billion in jobs, with a total economic impact of $3.6 billion. This is expected to increase with Northrop Grumman's future Ground Based Strategic Deterrent system.

A unique part of Hill Air Force Base is the Hill Aerospace Museum, opened in 1987, which holds a collection of more than 90 military aircraft, missiles, and aerospace vehicles on the museum's grounds. Inside are two exhibition galleries with many historic exhibits.

One

OGDEN ORDNANCE DEPOT
1920–1945

The history of Hill Air Force Base began during the Roaring Twenties, when the Ogden Ordnance Reserve Depot was created and designated a storage site for excess ammunition (ordnance) produced during World War I. In 1920, the federal government purchased 1,410 acres of land for the depot, with construction continuing through 1923. Starting that year and continuing through the Great Depression, the Ogden Reserve was used as a general US Army storage depot following reduced military funding. In 1927, its designation was changed to the Ogden Ordnance Depot. In 1936, funds were allocated for its rehabilitation and reconstruction, which included expansion for the manufacture and storage of bombs, shells, and other ordnance. Construction of new buildings and ammunition storage magazines was part of this expansion. In 1938, the facility began the production of munitions and construction of an air depot to meet US Army Air Corps preparations for possible US involvement in the war in Europe. The Japanese attack on December 7, 1941, at Pearl Harbor changed the depot's operations. The United States rapidly moved from peacetime to all-out support to fight. The depot stored military supplies that had been designated for the Philippines. By 1942, Ogden stored combat vehicles, with a wartime workforce of 6,000. The depot repaired different types of military aircraft starting in 1942. During the war, production expanded again to meet the large demands of a "two ocean war." From 1943 to 1946, the depot was the location of an Italian, and later German prisoner-of-war camp.

The Ogden Ordnance Reserve Depot was established in 1920. This aerial photograph shows the former arsenal administration building constructed in 1936 as part of the rehabilitation and reconstruction of the Ogden Arsenal. (Courtesy 75th Air Base Wing historian.)

This is the former arsenal administration building, where the Ogden Air Logistics Center Headquarters were until 2012, when it moved into the former headquarters building of the 309th Maintenance Wing. It had been used by the 75th Air Base Wing, 2849th Air Base Group, and 649th Air Base Group. The building remains after more than 80 years as a historic reminder of the continued military presence in the Salt Lake City area. (Courtesy 75th Air Base Wing historian.)

The screen and blend building for the black powder plant, seen here on July 9, 1941, was isolated from other buildings to prevent their destruction if it was torn apart by an accidental explosion. (Courtesy 75th Air Base Wing historian.)

Magazine No. 3026 stored bombs at the arsenal, served by a railroad spur track. The magazine was equipped with two doors to accept new munitions and load them into railroad boxcars using the wooden platform at the height of a boxcar's floor. In the distance, other similar structures are visible, separated to prevent the spread of damage from accidental explosions. This photograph was taken during the Historic American Buildings Survey of August 4, 2014. (Courtesy National Parks Service.)

Seen here are two views of a concrete ammunition storage igloo, with concrete facing, steel vault doors, and concrete unloading and loading pad. The storage igloo is covered with earth and thick sod to prevent erosion on the sloped sides. A power line is visible in the background. The igloos are separated to prevent damage from accidental explosions. These photographs were taken during the Historic American Buildings Survey of 2014. (Both, courtesy National Parks Service.)

Construction of the Ogden Ordnance Depot is in progress in this 1936 photograph. A utility trench is being hand-dug without safety walls, as would be required today. Depot warehouses are visible on the left. (Courtesy Utah State Historical Society.)

On February 1, 1934, the US Army Air Corps was charged with carrying airmail. One aircraft used on Airmail Route 18 through Salt Lake City was an open cockpit, World War I–era DeHavilland DH-4. It is on display at the National Museum of the United States Air Force. (Author's collection.)

A more advanced US Army Air Corps biplane, the P-12E was flown on Airmail Route 18. It also had an open cockpit, and flights through the mountains on this route required the pilots to wear heavy clothes. (Courtesy US Air Force.)

This photograph shows a twin-engine Douglas B-7 carrying the marking "AMROUTE 18" on its rear fuselage. (Courtesy Ogden Air Logistics Center, Hill Air Force Base.)

US Army Air Corps major Pete Hill was killed at Wright Field in Dayton, Ohio, on October 30, 1935, in a Boeing XB-17 prototype. On December 1, 1939, the Ogden air depot was renamed Hill Field. (Courtesy 75th Air Base Wing historian.)

A Boeing XB-17 prototype is being prepared for a test flight on the service ramp in front of a hangar on October 30, 1935. This was the US Army Air Corps' first four-engine, long-range strategic bomber to replace its smaller and less capable twin-engine bombers. (Courtesy National Museum of the US Air Force.)

On October 30, 1935, Maj. Pete Hill took off from Wright Field in this XB-17, crashing off the end of the runway, killing the major and two others. The photograph shows the XB-17 burning, with the four engines and front section of the fuselage destroyed. (Courtesy National Museum of the US Air Force.)

A ground-breaking ceremony was held on January 12, 1940, at Hill Field, with over 200 in attendance. An earth scraper and its tractor are behind the officials to create a background for this photograph. Earth leveling and runway preparations had already started. Construction was completed on September 1, 1941. (Courtesy 75th Air Base Wing historian.)

This is an October 1940 aerial photograph of Hill Field runway construction, which started on November 26, 1938. The runways were 7,500 feet long and 150 feet wide. This construction was part of Air Corps preparations for possible overseas combat as the military threat from Germany, Italy, and Japan grew. (Courtesy 75th Air Base Wing historian.)

This is a November 1940 photograph of Hill Field's runways, with hangars and support buildings at right center. (Courtesy 75th Air Base Wing historian.)

One of the first permanent buildings constructed at the air depot was the quartermaster commissary and warehouse, seen in this December 6, 1940, photograph. (Courtesy Ogden Air Logistics Center, Hill Air Force Base.)

Construction was completed on December 6, 1940, on the quartermaster garage and shop building, as seen here. (Courtesy Ogden Air Logistics Center, Hill Air Force Base.)

The fire guard and communications building is seen here on July 21, 1941. The three-bay firehouse is on the right side of the building. (Courtesy Ogden Air Logistics Center, Hill Air Force Base.)

This large multi-story barracks with attached post exchange was photographed on September 14, 1942. (Courtesy Ogden Air Logistics Center, Hill Air Force Base.)

This is an October 1942 photograph of completed hangar No. 1, with construction continuing on the adjacent hangar. (Courtesy 75th Air Base Wing historian.)

Three completed hangars are seen here with three Martin B-26 Invaders parked on the ramp. (Courtesy 75th Air Base Wing historian.)

Temporary office and storage space was initially provided by Work Progress Administration two-story barracks moved during April–June 1941 from the adjacent Ogden Ordnance Depot. (Courtesy Ogden Air Logistics Center, Hill Air Force Base.)

Several completed or nearly completed buildings are seen here, with the parking lot filled with civilian workers' prewar vehicles. (Courtesy Ogden Air Logistics Center, Hill Air Force Base.)

Among the first permanent buildings constructed at Hill Field were the officers' and noncommissioned officers' family quarters, seen in this July 18, 1941, photograph. (Courtesy Ogden Air Logistics Center, Hill Air Force Base.)

Construction started on the engine test facility on January 16, 1942. This August 31, 1942, photograph shows a completed building, with four test cells available. Operations were not yet in progress because the equipment had not been installed. (Courtesy Ogden Air Logistics Center, Hill Air Force Base.)

This is an April 6, 1942, photograph of the engine repair shop. After the United States entered World War II, US automakers shifted production to military equipment. All of these cars would have been built before the war. (Courtesy Ogden Air Logistics Center, Hill Air Force Base.)

A second engine test building was constructed to meet World War II demand, pictured in this August 31, 1942 photograph. Operations began on April 5, 1942, with the test of an R-2600-5 engine. (Courtesy Ogden Air Logistics Center, Hill Air Force Base.)

Here is a side view of Ogden Air Logistics Center Headquarters, listed as building No. 2, shown in 1940. Construction equipment and piles of building supplies are along the side of the building. (Courtesy Ogden Air Logistics Center, Hill Air Force Base.)

This July 14, 1944, photograph shows one of several supply division warehouses constructed in the Hill Field lower area and used for storage. (Courtesy Ogden Air Logistics Center, Hill Air Force Base.)

This aerial photograph of Hill Field was taken in October 1943, with runways, taxiways, hangars, and support buildings visible. (Courtesy 75th Air Base Wing historian.)

This aerial photograph taken from an altitude of 16,000 feet on October 3, 1943, shows completed buildings and the runway pattern. (Courtesy Ogden Air Logistics Center, Hill Air Force Base.)

This December 1940 photograph shows progress on warehouses and the operations hangar in the background, with its roof trusses being erected. The warehouses would be served by a railroad spur for loading and unloading. (Courtesy Ogden Air Logistics Center, Hill Air Force Base.)

Roof trusses are in place on a warehouse built of wood with steel joint supports to decrease the amount of scarce high-priority steel used for military equipment. (Courtesy 75th Air Base Wing historian.)

This large hangar was constructed of brick and steel, unusual for wartime projects when wood was the main construction material. This is a March 1943 photograph of the depot maintenance hangar under construction. (Courtesy Ogden Air Logistics Center, Hill Air Force Base.)

Hill Air Force Base renovated these two World War II maintenance hangars, which remained functional after nearly 80 years. (Courtesy 75th Air Base Wing historian.)

A restored World War II Boeing B-17G Flying Fortress with its distinctive chin-mounted twin .50 caliber machine gun turret flies low over Hill Air Force Base. (Courtesy 75th Air Base Wing Public Affairs.)

This building was used as the ordnance repair and reclamation workshop. This photograph was taken during the Historic American Buildings Survey in 2014. (Courtesy National Parks Service.)

This is the entrance to the black powder screening and blending building used in the production of 37mm anti-tank ammunition primer. This image was captured during the Historic American Buildings Survey of 2014. (Courtesy National Parks Service.)

Pictured is the warehouse used for bin storage to support the retail and wholesale issue of general military supplies. This photograph was taken during the Historic American Buildings Survey in 2014. (Courtesy National Parks Service.)

After the Japanese attack on Pearl Harbor on December 7, 1941, the US Army Air Corps transferred these Consolidated B-24 Liberators to Hill Field from West Coast airfields as a precaution. (Courtesy Ogden Air Logistics Center, Hill Air Force Base.)

A continuous line of Consolidated B-24 Liberators is on the progressive line in the repair hangar. The photograph shows the 1943 production line, the first such US Army Air Forces operation during World War II. (Courtesy 75th Air Base Wing historian.)

The first Boeing B-17E Flying Fortress to complete a round-the-world flight, named *Suzy-Q*, arrived at Hill Field on July 19, 1943, for overhaul. The B-17E left Hill on September 23, 1943. (Courtesy Ogden Air Logistics Center, Hill Air Force Base.)

This July 1944 photograph shows the outdoor storage of Consolidated B-24 Liberator bombers and Republic B-47 Thunderbolt fighters. (Courtesy 75th Air Base Wing historian.)

Military and civilian personnel gather on the concrete parking apron in front of the repair hangar on June 5, 1944, to celebrate the successful Allied invasion of Normandy on D-Day, the day before. (Courtesy 75th Air Base Wing historian.)

On May 15, 1943, under the jurisdiction of the US Army, Hill Field became the site of a prisoner-of-war camp, with an initial assignment of 724 Italian prisoners captured in North Africa, reaching a total of 823. (Courtesy US Army Signal Corps.)

In September 1944, the Italian government signed an armistice with the Allies, having declared war on Germany in June. Italian service troops worked on the base, as shown here loading ammunition into a railroad boxcar. (Courtesy US Army Signal Corps.)

German prisoners of war replaced the Italian prisoners at Hill Field, although in lower numbers. Even though Germany surrendered on May 7, 1945, German prisoners remained in the camp until June 1946. (Courtesy US Army Signal Corps.)

This post–World War II aerial photograph shows Hill Field with nearly all outdoor storage ground full of surplus military aircraft. (Courtesy 75th Air Base Wing historian.)

In this 1946 photograph, World War II PT-13 Stearman biplane trainers, no longer needed to train pilots, are seen stacked nose down to increase storage capacity. (Courtesy Ogden Air Logistics Center, Hill Air Force Base.)

Two

UTAH TEST AND TRAINING RANGE

1942–PRESENT

The desert military range was created in 1940 when Congress appropriated funds for the acquisition of land for bombing and gunnery ranges. In 1941, the US Army Air Corps received 1.8 million acres of land to support a bombing and gunnery range detachment on Wendover Army Air Base. In 1942, this facility trained Boeing B-17 Flying Fortress and Consolidated B-24 aircrews. In 1942, 126,720 acres were transferred from public domain to the war department to create the Dugway Proving Grounds. In 1944, Republic P-47s used the Wendover Range, followed by the 509th Composite Group. In 1945, Wendover Army Air Base was transferred to the Ogden Air Technical Command, designated the Ogden Air Logistics Complex, which continued to be used as a bombing and training range. In 1955, the Air Force assigned its air munition functions to the Ogden Air Logistics Complex. In 1960, the Wendover Range was designated Hill Air Force Range. Construction of a munitions and missile test facility at Oasis on the North Range was completed in 1964. In the early 1970s, an air-to-ground scoreable gunnery range was constructed to train fighter and bomber aircrews. The Air Force Systems Command assumed control of the range in 1979, and redesignated the Wendover/Hill/Dugway Range as the Utah Test and Training Range. Its mission is to test and evaluate cruise missiles and provide weapons testing and support. In 1997, Air Combat Command assumed responsibility for the range as a result of the 1995 base realignment and closure process. The 388th Range Squadron carried out the range's mission to provide open-air training and testing to support day-to-day training, large force training exercises, and large footprint weapons testing, and to serve as daily manager of range scheduling. The range averages approximately 16,000 training and 300 test flights each year. The highest usage was in 1994, with 22,229 sortie operations.

For years after World War II, the Utah Test and Training Range was called the Hill Air Force Base Range, administered by the Ogden Air Logistics Center, currently under the 388th Fighter Wing. This is a photograph of the warning sign. (Courtesy 75th Air Base Wing historian.)

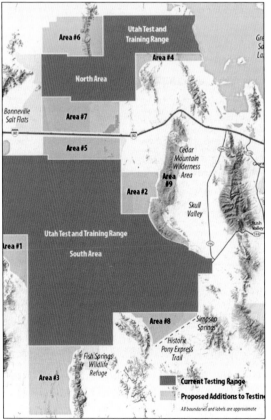

The Utah Test and Training Range comprises 12,574 square miles, of which there are 6,010 square miles of restricted air space and 6,564 square miles of military operating areas. The Department of Defense controls 2,624 square miles, of which the Air Force owns 1,490 square miles, with the US Army at Dugway Proving Grounds owning the balance. (Courtesy US Air Force.)

This is a warning sign on the Utah Test and Training Range, which is used for air-to-air and air-to-ground simulated combat, along with inert and live practice bombing and gunnery training. (Courtesy US Air Force.)

The Wendover Bombing and Gunnery Range began in 1942, training Boeing B-17 Flying Fortress and Consolidated B-24 aircrews based at Wendover Army Air Base. In 1944, Republic P-47 Thunderbolt fighter pilots trained on the base. This was followed by Boeing B-29 Superfortress aircrews from the 509th Composite Group. (Courtesy US Air Force.)

This is an aerial photograph of post–World War II Wendover Army Air Base with the rusting B-29 *Enola Gay* hangar and surrounding wooden structures. According to Col. Paul Tibbets, commander of the 509th Composite Group, "Wendover's runways could accommodate the Superfortress, with hangar and maintenance facilities in relatively good shape in September 1944." (Courtesy Historic Wendover Airfield.)

This is a 1995 photograph of World War II Wendover Army Air Base wooden barracks, rapidly built and designed to last only five to seven years (the estimated length of World War II), used to house B-17 and B-24 aircrews, and the 509th Composite Air Group, which dropped test bombs resembling the two atomic bombs dropped on Japan. Colonel Tibbets said, "When we inspected the wood barracks, they were substandard but good enough for the short time we were expected to be on the base." (Courtesy Historic Wendover Airfield.)

This 1942 photograph of Wendover Army Air Base shows the many different wooden buildings used for bombardment group training, which included high-altitude bombing practice with bombardiers using the top-secret Norden bombsight. (Courtesy Historic Wendover Airfield.)

Seen here is a 1945 photograph of the 509th Composite Group, 393rd Bomb Squadron hangar, now referred to as the "Enola Gay hangar." It was used for testing and modifications of the B-29 used for dropping dummy test bombs to determine ballistics and accuracy before the actual atomic bombs were dropped on Japan. (Courtesy Historic Wendover Airfield.)

This is a 1995 photograph of the World War II *Enola Gay* hangar in disrepair, with extensive metal siding and roofing rust after more than 50 years since its construction. (Author's collection.)

The renovated *Enola Gay* hangar is pictured here in 2020. The Historic Wendover Airfield Foundation received a generous grant from the State of Utah to continue the hangar's restoration. The hangar's doors were taken down and straightened, and the bent members were replaced. The doors can still be manually opened, as during World War II. The hangar's exterior was painted. (Courtesy Historic Wendover Airfield.)

These earth-covered concrete munitions storage igloos on the former Wendover Army Air Base were used to store bombs and ammunition expended on the Wendover Bombing and Gunnery Range. According to Col. Paul Tibbets, "The group's bombardiers' task was to drop atomic dummy ballistic test bombs onto a bombing range into a ground target that consisted of a circle 400 feet in diameter. Our desired accuracy was to place the bombs no more than 200 feet from the aiming point from an altitude of 30,000 feet." (Courtesy Historic Wendover Airfield.)

Shown are members of the Special Ordnance Division, 216th Army Air Force Ballistic Unit, which assembled inert bombs of various ballistic shapes. Here, they are lowering a test bomb, nicknamed a "pumpkin," into a bomb loading pit. The Wendover Bombing and Gunnery Range allowed ballistic testing, fusing mechanism testing, and release mechanism modifications. (Courtesy Historic Wendover Airfield.)

Pictured is an inert ballistic test bomb that was a copy of the "Fat Man" plutonium atomic bomb. The 509th Composite Group dropped approximately 200 "pumpkin" test bombs during secret testing on the isolated and secure Wendover Bombing and Gunnery Range. Accuracy from 30,000 feet was required. (Courtesy Historic Wendover Airfield.)

Concrete bomb loading pits were constructed on Wendover Army Air Base, with sufficient dimensions to hold either the "Little Boy" or Fat Man atomic bombs. The inert test bombs were lowered into the pits on top of a hydraulic lift platform, which would lift the bombs up into a Boeing B-29 Superfortress bomb bay. The dimensions of the pits were sent to engineers on North Field, Tinian Island, so they could build duplicates there. (Courtesy Historic Wendover Airfield.)

A 509th Composite Group Boeing B-29 Superfortress is over a Wendover atomic bomb loading pit with a "pumpkin" ballistic test bomb being lifted into the bomb bay. The bomb was painted black and white to allow photographic study of its movements from 30,000 feet down to the target in the Wendover Bombing and Gunnery Range. (Courtesy Historic Wendover Airfield.)

This 2019 photograph was taken at the Utah Test and Training Range of a bomb exploding inside a selected target circle as during World War II, when the 509th Composite Group dropped ballistic text bombs into a circle marked on the desert floor of the Wendover Bombing and Gunnery Range. (Courtesy US Air Force.)

The Historic Wendover Airfield staff recovered these rusted inert training bombs from the Utah Test and Bombing Range area. (Courtesy Historic Wendover Airfield.)

On May 29, 1945, the 509th Composite Group's Boeing B-29 Superfortresses landed on North Field, Tinian Island, after a flight from Wendover Army Air Base. Pictured is the group's headquarters area built on North Field by US Navy Construction Battalion men, one of whom was the author's father, George W. Larson. (Author's collection.)

This aerial photograph taken over North Field, Tinian Island, on April 24, 1987, shows the two atomic bomb loading pits at left. The pits are set apart from the runways in a Y-shaped parking ramp, which contains pit No. 1 at the bottom (used to load the Little Boy bomb) and pit No. 2 at the top (used to load the Fat Man bomb). (Author's collection.)

This is a duplicate of the Little Boy atomic bomb. Its ballistic characteristics and systems were tested at the Wendover Bombing and Gunnery Range. Range operations were supported by Hill Field. The replica is on display at the Historic Wendover Airfield Museum. (Courtesy Historic Wendover Airfield.)

Above, on August 5, 1945, at North Field on Tinian Island, the 509th Composite Group Boeing B-29 Superfortress *Enola Gay* is slowly backed over bomb loading pit No. 1. This pit was constructed to replicate the bomb loading pit on Wendover Army Air Base. Below, the Little Boy atomic bomb is being lifted out of loading pit No. 1 into the bomb bay of the *Enola Gay*. These images were taken by the 509th Composite Group photographer. (Author's collection.)

This is a January 1981 photograph of atomic bomb loading pit No. 1, northwest of North Field runway No. 4. After World War II, it fell into disrepair and eventually filled with dirt and tropical plants. (Author's collection.)

This is a January 1981 photograph of the bronze memorial plaque mounted on top of a concrete pedestal in front of loading pit No. 1. (Author's collection.)

Tinian's residents eventually dug out pit No. 1 and fabricated a steel and glass covering over both bomb loading pits to protect them from the elements. (Author's collection.)

The Fat Man plutonium bomb is on its transporter, positioned over loading pit No. 2. The bomb was covered with a tarp to protect it from rain squalls on the island. This photograph was taken by John Vandervort, 509th Composite Group photographer. (Author's collection.)

A Fat Man replica is on display at the National Museum of the US Air Force. Approximately 200 inert "pumpkin" ballistic bombs were tested by the 509th Composite Group to determine release, impact, and detonation specifications on the Wendover Bombing and Gunnery Range. (Author's collection.)

The Fat Man atomic bomb is partially lowered into loading pit No. 2 on August 8, 1945. It was then loaded into the bomb bay of a 509th Composite Group B-29 named *Bockscar*. This photograph was taken by John Vandervort, 509th Composite Group photographer. (Author's collection.)

This is the entrance to the Utah Test and Training Range, South Area, in northern Utah, west of Salt Lake City and 15 miles south of Interstate 80. The test and range areas include a north and south area. The South Area includes the World War II Wendover Bombing and Gunnery Range, with over a million acres of secured land used continuously since 1942, as well as the Dugway Proving Ground and the Army's chemical and biological testing area. (Courtesy Center of Land Use Interpretation.)

This 2016 photograph shows the destruction of a Trident II rocket motor. The destruction of three-stage, solid propellant rocket motors continues for the Navy's Trident and Air Force's Minuteman missiles to eliminate aged propellants and to meet treaty obligations. (Courtesy US Air Force.)

This is an unmanned AGM-86B air-launched cruise missile over the Utah Test and Training Range approaching its target during a nuclear weapons system evaluation program simulated combat mission on September 22, 2014. The missile was launched from a Barksdale Air Force Base 2nd Bomb Wing Boeing B-52H Stratofortress bomber as part of an end-to-end operational evaluation of 8th Air Force Task Force 204's capability to deliver weapons from storage to a target. (Courtesy Air Material Command Public Affairs.)

This image was taken during the first 388th Fighter Wing General Dynamics F-16 Fighting Falcon strafing evaluation mission on the Utah Test and Training Range. (Courtesy 75th Air Base Wing Public Affairs.)

Shown is a modified Air Force Reserve Lockheed C-130 Hercules assigned to the 910th Airlift Wing from Youngstown Air Reserve Station, Ohio, conducting aerial spraying of herbicide on selected areas of the Utah Test and Training Range on March 18, 2015, to eliminate invasive weeds on the range. (Courtesy Youngstown Air Reserve Station Public Affairs.)

General Dynamics F-16 Fighting Falcon maintainers from Shaw Air Force Base, South Carolina, are on the Hill Air Force Base flight line on May 6, 2014. F-16s participated in an Air Force weapon system evaluation program during air-to-ground exercise Combat Hammer. F-16s dropped lived munitions in realistic combat scenarios at the Utah Test and Training Range. (Courtesy 75th Air Base Wing Public Affairs.)

A 28th Bomb Wing B-1B Lancer bomber is shown landing on May 15, 2012, on Ellsworth Air Force Base, South Dakota, after participating in exercise Combat Hammer over the Utah Test and Training Range. During the exercise on May 14–15, 2012, B-1Bs employed GBU-54 Joint Direct Attack Munitions against moving targets. (Courtesy 28th Bomb Wing Public Affairs.)

A Lockheed B-2 stealth bomber is shown flying over the Utah Testing and Training Range on August 27, 2003, dropping a string of inert Joint Direct Attack Munitions. (Courtesy US Air Force.)

A General Dynamics F-16 Fighting Falcon assigned to the 79th Fighter Squadron fires the AGM-65D Maverick air-to-ground missile over the Utah Test and Training Range during exercise Combat Hammer on August 7, 2002. (Courtesy Air Material Command Public Affairs.)

A General Dynamics F-16 Fighting Falcon, flown by the 160th Fighter Squadron, 187th Fighter Wing of the Alabama National Guard, drops a GBU-24A laser-guided bomb over the Utah Test and Training Range on July 2, 2002, during Combat Hammer. (Courtesy Alabama Air National Guard Public Affairs.)

The Republic F-105 Thunderchief was assigned to the 419th Fighter Wing. The F-105 was designed as a nuclear bomber, becoming an outstanding heavy conventional bomber during the Vietnam War. Shown are three F-105s in formation over the Utah Test and Training Range. (Courtesy 75th Air Base Wing historian.)

Shown are airmen assigned to the 23rd Aircraft Maintenance Squadron, Moody Air Force Base, Georgia, beginning the offload of 30mm shells from a Republic Thunderbolt II on August 3, 2016, after completing a mission during exercise Combat Hammer at Hill Air Force Base. (Courtesy 75th Air Base Wing historian.)

These three Lockheed Martin F-22 Raptors from the 1st Fighter Wing assigned to Langley Air Force Base, Virginia, are pictured in formation over the Utah Test and Training Range during a weapons system evaluation program on August 12–23, 2007, under exercise Combat Hammer. (Courtesy 75th Air Base Wing Public Affairs.)

On May 30, 2013, the Air Force transferred Lockheed Martin F-22 Raptor depot maintenance to Hill Air Force Base, the Ogden Air Logistic Complex. Shown is an F-22 under the open aircraft shelter. The aircraft, once completing depot maintenance, has the nearby Utah Bombing and Test Range for post-maintenance test flights and weapons testing. (Courtesy 75th Air Base Wing Public Affairs.)

Three

POST–WORLD WAR II
1946–1970s

The postwar years brought changes, starting on March 16, 1947, with the Army Air Forces transferring jurisdiction of Wendover Army Air Base from Ogden Air Material Area to the Strategic Air Command. Aircraft storage was a major operation in 1947. On September 18, the Army Air Forces became the US Air Force. On July 2, 1948, Ogden Air Material Area assisted in supporting the Berlin Airlift to break the Russian blockade of West Berlin. On June 25, 1950, the invasion of South Korea by North Korean military gave the Ogden Air Material Area an emergency opportunity to implement logistic support of United Nations forces fighting in Korea. On December 13, 1950, the first Boeing B-29 Superfortress entered maintenance for transfer to the Far East Air Force in Korea. The B-26 production line started on October 1, 1950, preparing the World War II aircraft for a ground-support role in Korea. In 1955, the Ogden Arsenal was transferred to Hill Air Force Base. The base completed a 13,500-foot runway modernization to improve air operations. In August 1959, the Ogden Air Material Area implemented a time compliance technical order, which called for a specific aircraft work schedule for required maintenance or modifications. On November 26, 1956, the Air Force was assigned exclusive combat control over all missiles with a range in excess of 200 miles. On January 6, 1959, the Air Material Area was assigned management responsibilities for the Minuteman intercontinental ballistic missile. The Hill Air Force Range was part of Hill Air Force Base, with construction beginning on October 2, 1962, and dedicated on July 31, 1964. Hill logistics supported the war in Southeast Asia beginning in 1966, especially with ammunition flown on Military Airlift Command C-124s from the base beginning on March 3, 1966. On April 1, 1976, the Ogden Air Material Area was redesignated the Ogden Logistic Center. The center was an important Air Force missile support center, including the Strategic Air Command's upgrade to the Minuteman III.

This aerial photograph taken in May 1948 shows the large number of World War II surplus aircraft parked in the outdoor storage area of Hill Air Force Base. Aircraft were parked in an environment that was not conducive to permanent storage. (Courtesy 75th Air Base Wing historian.)

To speed production of 42 Boeing B-29 Superfortresses, Hill Air Force Base maintenance developed and implemented a continuous maintenance line. This photograph shows B-29s that were removed from outdoor storage, beginning maintenance work on May 4, 1950. (Courtesy 75th Air Base Wing historian.)

Pictured are B-29s inside the maintenance hangar for Korean War maintenance and modification. After North Korea invaded South Korea on June 25, 1950, World War II B-29s were sent to the Far East Air Force for combat operations, wading into air attacks by Soviet-supplied MiG-15 fighters. (Courtesy 75th Air Base Wing historian.)

This is a 1953 aerial photograph of Hill Air Force Base operations building (hangar No. 1) with the runway control tower, which was at that time mounted on top of the building. More maintenance hangars are visible in the background. (Courtesy 75th Air Base Wing historian.)

For a brief time, the 3005th, formerly World War II Women in the Air Corps (WACs), was assigned to Hill Field from 1943 to 1947; later, Air Force–designated Women in the Air Force (WAFs) were assigned to the base on May 16, 1951. The squadron remained on the base until December 6, 1954. (Courtesy 75th Air Base Wing historian.)

This Hill Air Force Base World War II–constructed chapel, photographed on March 20, 1951, used for Protestant, Catholic and Jewish services, was moved to the Hill Aerospace Museum. (Author's collection.)

Northrop P-61 Black Widow night fighters and other aircraft were processed in 1947. This photograph is of the Northrop P-61 Black Widow repair line in the maintenance hangar, with a Boeing B-29 Superfortress partially visible in the background, outside the hangar. (Courtesy 75th Air Base Wing historian.)

This is a June 1958 photograph of the Hill Air Force Base entrance into the Ogden Air Material Area. (Courtesy 75th Air Base Wing historian.)

Northrop F-89 Scorpion production takes place in this June 30, 1958, photograph taken in aircraft repair hangar Nos. 3 and 4. Maintenance personnel followed the time compliance technical orders for this aircraft to complete required maintenance work. (Courtesy 75th Air Base Wing historian.)

This public affairs photograph taken on the Hill Air Force Base flight line celebrates the Ogden Logistic Center's first Republic F-84 Thunderjet reconditioning. (Courtesy 75th Air Base Wing historian.)

This is the first Republic F-84F swept-wing Thunderstreak on the maintenance line in May 1955. (Courtesy 75th Air Base Wing historian.)

In this July 1950 photograph, maintenance personnel climb all over a Douglas B-26 Invader pulled from outdoor storage for preparation for shipment to the Far East Air Force for ground combat in Korea to support United Nations troops fighting the North Korean troops. (Courtesy 75th Air Base Wing historian.)

Shown is the last of approximately 2,000 B-26 Invaders that were processed through the production line on August 27, 1955, closing an eight-year project. (Courtesy 75th Air Base Wing historian.)

This is an aerial view of Capehart housing duplexes on Hill Air Force Base in the early 1960s, consisting of three to four bedrooms and split-level and ranch-style architecture with construction of brick, wood siding, stucco, and other finishes. (Courtesy 75th Air Base Wing historian.)

This is a 1957 aerial view of the completed 13,500-foot-long and 200-foot-wide runway, dedicated on September 30, 1956, and completed on March 4, 1957. (Courtesy 75th Air Base Wing historian.)

Seen inside the maintenance hangar on Hill Air Force Base are the maintenance lines for the swept-wing Republic F-84F Thunderstreaks, crowded with an almost endless line of aircraft undergoing maintenance. (Courtesy 75th Air Base Wing historian.)

This image was taken on June 10, 1959, of a McDonnell F-101 Voodoo photo-reconnaissance variant being processed under the time compliance technical order in the maintenance hangar. (Courtesy 75th Air Base Wing historian.)

This is a July 1, 1969, photograph of Convair F-102 Delta Dagger supersonic fighter-interceptors angled with their noses toward the hangar's walls on the production lines in hangar Nos. 3 and 4. (Courtesy 75th Air Base Wing historian.)

A Convair F-102 Delta Dagger is seen here at the Hill Aerospace Museum. It was the Air Force's first delta-wing, supersonic fighter-interceptor. It was only armed with air-to-air missiles—no guns. The Ogden Air Material Area became a specialized repair maintenance facility for the F-102 in 1957. (Author's collection.)

The Ogden Air Material Area set up the first production maintenance line for the Northrop F-89 Scorpion fighter-interceptor. This F-89 is on display at the Hill Aerospace Museum. The production line began on January 29, 1953. (Author's collection.)

This restored Douglas A-26 Invader is on display inside the Hill Aerospace Museum. (Author's collection.)

The Ogden Air Material Area repaired 37 Martin B-57 Canberra Night Intruder bombers. A B-57 is on display at the Hill Aerospace Museum. From 1953 to 1956, the 451st Bombardment Wing flew the B-57 at Hill Air Force Base. (Author's collection.)

In April 1956, the Ogden Air Material area gained procurement responsibilities for the Air Force's T-38 Talon two-seat supersonic trainer. This aircraft is on display at the Hill Aerospace Museum. (Author's collection.)

The Republic F-84 Thunderstreak was phased out of Hill Air Force Base and Ogden Air Material Area Maintenance on May 26, 1958, with all such work relinquished on July 1, 1959. An F-84 is shown on display at the Hill Aerospace Museum. (Author's collection.)

The new base control tower is under construction in 1973; it was completed in 1974. The control tower's observation dome is being raised for placement at the top of the concrete base by a heavy lift crane. (Courtesy 75th Air Base Wing historian.)

Two Aircraft Repair Facility hangars are seen here at Hill Air Force Base, with two more hangars behind them. (Courtesy 75th Air Base Wing historian.)

Four

Ogden Air Logistics Center/Complex

1970–Present

In the 1970s, there were serious funding cuts in Department of Defense agencies with the end of the Vietnam War. On January 1, 1973, Detachment 1, 466th Bombardment Wing with B-52s were stationed at Hill Air Force Base on satellite nuclear mission in alert facilities. Air Force Reserve units were active on the base in the 1970s, with the Republic F-105 Thunderchiefs assigned on November 27, 1972. On January 1, 1979, management of the Hill/Wendover/Dugway Range complex was transferred to Air Force System Command, renamed the Utah Test and Training Range. On December 26, 1976, the Air Logistics Center became the worldwide system and maintenance management facility for the General Dynamics F-16 Fighting Falcon. Hill Air Force Base was an important missile production center. This included the April 7, 1980, start of the MX missile logistics management and support, but the missile was eventually pulled from silo alert at F.E. Warren Air Force Base. The 388th Tactical Fighter Wing received its first F-16 on January 6, 1979. The Logistics Center continued to support F-101s, F-4s, and F-16s worldwide in the early 1980s. The 388th and 419th Tactical Fighter Wings used the Utah Test and Training Range, flying against other fighter wings in simulated air combat. Also during the 1980s, the Logistics Center remained the system program manager for the intercontinental ballistic missile deterrent. It was designated the Peacekeeper Rail Garrison program manager, but the deployment was canceled in December 1961. From 1990 to 1991, the Logistics Center and Hill Air Force Base tenant units supported Operations Desert Shield and Desert Storm. After September 11, 2001, the Logistics Center began a massive support effort for America's war on terrorism and continued maintenance on the Minuteman III, C-130, F-16, A-10, and F-22. On July 12, 2002, the Logistics Center became the Ogden Air Logistics Complex. In 2020, Hill Air Force Base employed over 20,000 military and civilian personnel with the Ogden Air Logistics Complex, 75th Air Base Wing, 388th and 499th Fighter Wings, and tenant organizations.

Maintenance workers finish work on one of the last Republic F-84s to undergo maintenance at Hill Air Force Base. (Courtesy 75th Air Base Wing historian.)

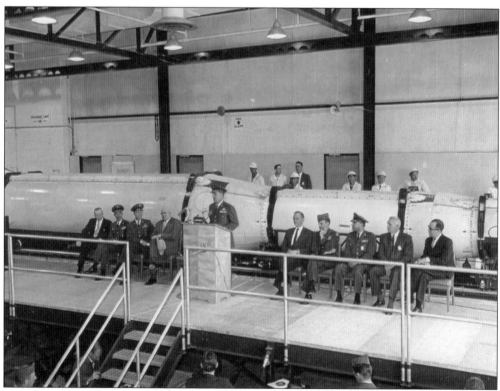

This is an April 12, 1962, photograph of the rollout ceremony for the first Minuteman I ICBM. (Courtesy 75th Air Base Wing historian.)

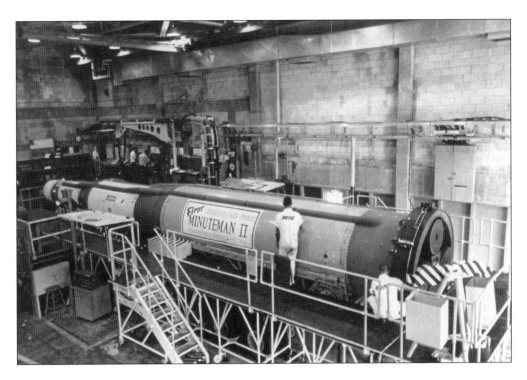

Above is the first Minuteman II ICBM at Hill Air Force Base; below is the assembly of the first advanced Minuteman II at the Boeing-operated Plant 77 at Hill. (Both, courtesy 75th Air Base Wing historian.)

The first operational Minuteman ICBM is being loaded into a C-133 Cargomaster for transport to Malmstrom Air Force base, to be one of the first of 10 missiles that were loaded into silos and placed on alert during the Cuban Missile Crisis in October 1962. (Courtesy 75th Air Base Wing historian.)

This C-133 Cargomaster was used to transport ammunition from Hill Air Force Base to South Vietnam to support the war in Southeast Asia. It had sufficient cargo capacity to transport Minuteman ICBMs to Strategic Air Command intercontinental ballistic wings throughout the United States. (Courtesy 75th Air Base Wing historian.)

In January 1973, the 456th Bombardment Wing, Detachment 1, Strategic Air Command, was activated as a satellite alert facility at Hill Air Force Base. On December 28, the first four Boeing B-52 Stratofortresses and two Boeing KC-135A Stratotankers assumed 24-hour nuclear deterrent alert. Nuclear alert was discontinued on July 1, 1975. (Courtesy 75th Air Base Wing historian.)

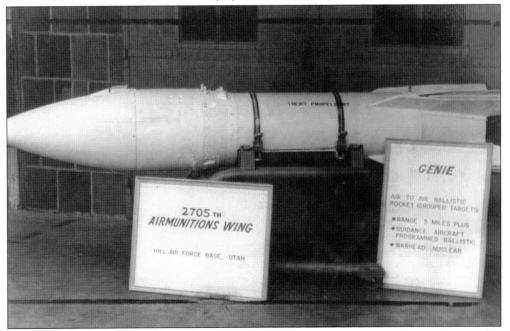

Pictured here is Genie, a nuclear warhead–armed, unguided, air-to-air anti-craft rocket designed to shoot down mass formations of penetrating Russian Air Force strategic bombers carrying nuclear weapons to attack the United States. (Courtesy 75th Air Base Wing historian.)

Here is a Fairchild/Republic A-10 Thunderbolt II ground-attack fighter undergoing maintenance inside an Ogden Air Logistic Complex maintenance hangar at Hill Air Force Base. (Courtesy 75th Air Base Wing historian.)

Technicians of the 576th Aircraft Maintenance Squadron conduct corrosion control by applying primer coating to an A-10 in preparation for final exterior painting. (Courtesy 75th Air Base Wing historian.)

An A-10 mechanic assigned to 571st Aircraft Maintenance Squadron is shown working on a Thunderbolt II's wing. (Courtesy 75th Air Base Wing historian.)

This is an A-10 in its final process of maintenance before the technicians complete all work. The Ogden Air Logistics Complex also supports maintenance and distribution on the F-35A, F-16, C-130, T-38, Minuteman III, and other weapon systems. (Courtesy 75th Air Base Wing historian.)

A Lockheed Martin F-35A Lightning II undergoes maintenance. (Courtesy 75th Air Base Wing historian.)

This is a Maverick air-to-ground optical guided cruciform wing weapon system. The first maintenance of this missile type was an overhaul of a guidance unit on October 31, 1972. (Courtesy Ogden Air Logistics Center, Hill Air Force Base.)

This is a mockup of an MX missile at Hill Air Force Base. On September 2, 1975, the Ogden Air Logistics Center was assigned management of this missile, scheduled to be the replacement for the Minuteman III ICBM. (Courtesy 75th Air Base Wing historian.)

This is a Northrup Grumman public affairs conceptual image of the proposed ground-based strategic deterrent ICBM to replace the Minuteman III. (Courtesy Northrup Grumman Public Affairs.)

Pictured is a Minuteman III with six missile maintenance technicians completing an upgrade on the missile with new cabling and external protective cable covering installed. (Courtesy 75th Air Base Wing historian.)

This is an IM-99 Bombarc surface-to-air anti-aircraft missile. By the end of 1958, the Ogden Air Material Area became the weapon storage site for this missile and its system management. (Courtesy Ogden Air Logistics Center, Hill Air Force Base.)

An elevated Bombarc missile is on a display trailer in front of repair hangar No. 3 at Hill Air Force Base. It was designed as a long-range surface-to-air missile to shoot down Russian bombers attacking the United States. (Courtesy 75th Air Base Wing historian.)

An Inert GAM-87 Skybolt is on its transporter alongside a Boeing B-52 Stratofortress bomber. (Courtesy US Air Force.)

Shown is a Boeing B-52 Stratofortress bomber carrying four GAM-87 Skybolt 40-foot-long, 1,000-mile-range ballistic missiles. Ogden Air Material Area was designated the missile's system manager on July 6, 1959. The missile was canceled and never operational. (Author's collection.)

Seen here is a 2015 photograph of a Hill Air Force Base runway, with munitions storage igloos to the right and an F-16 fighter ramp in the background. (Courtesy 75th Air Base Wing historian.)

Five

THE 388TH FIGHTER WING
1953–PRESENT

On March 23, 1953, the 388th Fighter Day Wing was activated. It was designated the 388th Fighter-Bomber Wing on November 5, 1953. It was inactivated in 1957, designated the 388th Tactical Fighter Wing on May 1, 1962, and equipped with the Republic F-105 Thunderchief, participating in the Vietnam War, and ending combat operations flying the McDonnell Douglas F-4 Phantom II. In April 1977, the wing moved to Hill Air Force Base, becoming the first Air Force wing equipped with the General Dynamics F-16 Fighting Falcon multi-role aircraft. During the initial stages of the conversion, the wing trained F-16 instructor pilots and provided replacement training for new F-16 pilots. In March 1981, the wing conducted its first overseas deployment to Fleshan Air Station, Norway. The 388th Tactical Fighter Wing was first to fly the F-16 into combat, equipped with the Low-Altitude Navigation and Targeting Infrared for Night System (LANTIRN) in Iraq and Kuwait during Operations Desert Shield and Desert Storm. Wing F-16s initially deployed to Spain as attrition reserves from January to December 1991. Two of the wing's squadrons deployed to Southwest Asia for combat operations from August 28, 1990, to March 27, 1991. After the March 1991 cease-fire, the wing remained to protect coalition assets and make certain Iraq compiled with treaty terms, beginning in December 1991. The 388th Fighter Wing continued to support LANTIRN-equipped F-16s in Operations Desert Calm and Desert Fox as well as Southern and Northern Watch. The 388th Fighter Wing and the 388th Range Squadron, coupled with the Utah Test and Training Range, have continued to provide test and training for the F-16 fleet through local and off-base exercises Air Warrior, Amalgam Thunder, Combat Archer, Cope Thunder, Iron Falcon, Maple Flag, and Red Flag and support for the fighter weapon instructor course and tactical air control parties. With the landing of the first Lockheed Martin F-35A Lightning II fighter at Hill, the phase-out of the F-16s began. By 2019, the wing's strength reached a total of 87 F-35As.

Seen here is the North American F-100D Super Sabre, flown by the 388th Fighter-Bomber Wing in 1957 and 1962–1964, when it was flown by the 388th Tactical Fighter Wing. The F-100 was the Air Force's first supersonic fighter, then fighter-bomber during the Vietnam War. This aircraft is on display at the South Dakota Air and Space Museum. (Author's collection.)

A Republic Aviation F-105 Thunderchief, flown by the 388th Fighter Wing in 1963–1964, 1966–1969, and 1970–1974, is seen here. It was the Air Force's largest, heaviest, supersonic, single-seat fighter-bomber of its time. It was heavily flown during the Vietnam War, suffering high losses. This aircraft is on display at the South Dakota Air and Space Museum. (Author's collection.)

Pictured here is the McDonnell Douglas F-4 Phantom II flown by the 388th Fighter Wing in 1968–1975 and 1976–1980. It was a two-seat, multi-role, supersonic fighter-bomber. The aircraft is on display at the Kansas Museum of Flight. (Author's collection.)

The Douglas B-66 Destroyer was designed as a tactical bomber, conventional and nuclear. One of its modifications was the EB-66 flown by the 388th as an electronic jammer, heavily used to confuse North Vietnamese radar in the air war over the north. This aircraft flown by the 388th from 1970 to 1974 is on display at the National Museum of the US Air Force. (Author's collection.)

The Lockheed EC-121R Warning Star was an airborne, early-warning aircraft that was used during the Vietnam War to detect and track North Vietnamese fighters supplied by the Soviet Union to shoot down attacking US aircraft. The Warning Star was flown by the 388th from 1970 to 1971. This aircraft is on display at Peterson Air Force Museum. (Author's collection.)

An EC-121R Warning Star undergoes restoration before display. (Author's collection.)

This Lockheed AC/130 Hercules, modified into a gunship, was flown by the 388th from 1972 to 1975 in the role of ground suppression of enemy positions with tremendous destructive firepower. (Courtesy US Air Force.)

Pictured is a Vought A-7D Corsair flown by the 388th from 1973 to 1975. The Air Force used the A-7D during the Vietnam War as an accurate ground-attack fighter-bomber. This aircraft is on display at the South Dakota Air and Space Museum. (Author's collection.)

This McDonnell Douglas F-4 Phantom II fighter-bomber was flown by the 388th from 1968 to 1975 and 1966 to 1969. The supersonic, twin-seat fighter-bomber was used to shoot down North Vietnamese fighters over North Vietnam. (Courtesy 75th Air Base Wing historian.)

A General Dynamics (now Lockheed) F-16 Fighting Falcon is parked on the 388th Fighter Wing flight line with a maintenance hangar in the background. The pilot is in the closed cockpit preparing for takeoff on a training flight. The aircraft's crew chief is to the right of the F-16. (Courtesy 75th Air Base Wing historian.)

Five F-16s from the 388th Fighter Wing, with squadron markings on the tail surfaces from the 421st, 34th, and 4th Squadrons, fly in line-abreast formation over Hill Air Force Base. (Courtesy 75th Air Base Wing historian.)

This is a view inside the maintenance hangar at Hill Air Force Base, where General Dynamics F-16 Fighting Falcons undergo phase maintenance. (Courtesy 75th Air Base Wing historian.)

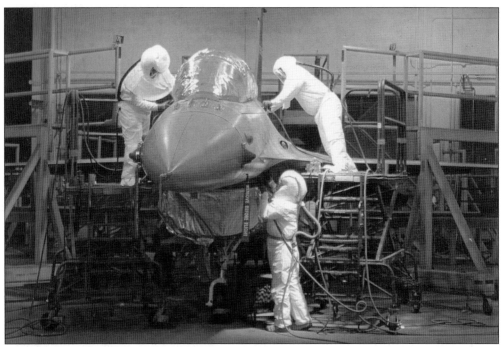

Maintenance personnel are pictured dressed in protective clothing with an outside air source pumped into their protective hoods to keep out harmful contaminants during corrosion maintenance on this F-16. The cockpit canopy is covered to prevent pitting of the clear canopy, which would degrade pilot visibility. (Courtesy 75th Air Base Wing historian.)

This is a 1995 photograph of a 421st Fighter Squadron F-16 on the Hill Air Force Base flight line, with three squadron hangars in the background. (Author's collection.)

This is a 1995 photograph of two 388th Fighter Wing F-16s on the Hill Air Force Base flight line. Shown is a two-seat version of the aircraft. (Author's collection.)

This is a view of 388th Fight Wing F-16s on the Hill Air Force Base flight line, with their hangars in the background. (Author's collection.)

Seen in this 1995 photograph is the 388th Fighter Wing flight line at Hill Air Force Base. (Author's collection.)

Three F-16s fly in formation past the mountains and into Hill Air Force Base. (Courtesy 75th Air Base Wing historian.)

These three F-16s from the 388th Fighter Wing were photographed from another 388th Fighter Wing F-16 flying slightly below and to the right. (Courtesy 75th Air Base Wing historian.)

These F-16C Fighting Falcons from the 388th Tactical Fighter Wing are on the flight line at an airfield in Southwest Asia during Operation Desert Storm on January 23, 1991. (Courtesy 388th Fighter Wing.)

Seen here is the North American F-100 Super Sabre, the Air Force's first supersonic fighter-interceptor. The 388th Fighter Wing flew the F-100 in 1957 and 1962–1964. An F-100 pilot is at the top of the access ladder, with the crew chief below. (Courtesy 75th Air Base Wing historian.)

This is an F-16 three-aircraft formation, with the lead aircraft marked with "Viper Out," commemorative tail marking flying with two other F-16s from the 388th Fighter Wing and 419th Fighter Wing on August 30, 2017. The Viper Out ceremony officially marked the end of 30 years of the F-16 at Hill Air Force Base. (Courtesy 388th Fighter Wing.)

Seen here is a dramatic head-on view of a Lockheed Martin F-35A Lightning II, one of the first two aircraft that landed at Hill Air Force Base on September 2, 2015. (Courtesy 75th Air Base Wing Public Affairs.)

One of the first two F-35A Lightning IIs that landed at Hill Air Force Base, assigned to the 34th Fighter Squadron, is being parked in the covered aircraft shelter. (Courtesy 75th Air Base Wing historian.)

This F-35A, flown by the 34th Fighter Squadron, is parked inside a hangar at Hill Air Force Base. The aircraft's two main landing-gear wheels are chocked to prevent movement and damage to the aircraft. (Courtesy 75th Air Base Wing historian.)

Shown is an F-35A of the 388th Fighter Wing positioned behind and slightly below a Boeing KC-135R Stratotanker, with the refueling receptacle open behind the cockpit and the tanker's refueling boom inserted for refueling. (Courtesy 75th Air Base Wing historian.)

A 388th Fighter Wing F-35A is pictured on a Hill Air Force Base flight line with two ground crew personnel approaching with wheel chocks to secure the aircraft. (Courtesy 75th Air Base Wing historian.)

A 388th Fighter Wing 34th Fighter Squadron F-35A is shown from an adjacent squadron F-35A. (Courtesy 75th Air Base Wing Public Affairs.)

Winter conditions can be extreme at Hill Air Force base, requiring large snow removal equipment to remove snow and ice from the runways, taxiways, and parking aprons to allow aircraft operations. The base control tower is visible in the background. (Courtesy 75th Air Base Wing historian.)

The 388th Fighter Wing flew many different fighters, such as this North American F-86F Sabre assigned to the 563rd Tactical Fighter Squadron in 1955. (Courtesy 75th Air Base Wing historian.)

The first two F-35As are seen above on a Hill Air Force Base taxiway on September 2, 2015, from the base control tower and below being directed to parking spots on the aircraft ramp. These were the first of the eventual 78 F-35As to be assigned to the 388th Fighter Wing. (Both, courtesy 75th Air Base Wing historian.)

Here is another view of the first two F-35As on September 2, 2015, on the flight line. (Courtesy 75th Air Base Wing historian.)

An F-35A begins a takeoff roll on a Hill Air Force Base runway. The base control tower is at left. (Courtesy 75th Air Base Wing historian.)

This is a November 19, 2018, photograph of 35 F-35As from the 388th and 419th Fighter Wings preparing for a possible massed takeoff in an exercise referred to as an "elephant walk." The aircraft were in close taxi formation to conduct a minimum-interval takeoff. The aircraft either took off or returned to their parking ramp. (Courtesy 75th Air Base Wing historian.)

A 388th Fighter Wing F-35A taxis to a refueling point during a surge exercise on September 9–13, 2019. The 34th and 421st Fighter Squadrons flew 240 sorties during this exercise. (Courtesy 75th Air Base Wing historian.)

A1c. Luigia Moriello, 388th Aircraft Maintenance Squadron, and Maj. James Schmidt (in the cockpit), assigned to the 388th Operations Group, exchange salutes during launch preparations at Hill Air Force Base during exercise Combat Hammer, part of the Weapons System Evaluation Program. (Courtesy US Air Force.)

An F-35A launches an AIM-120 missile at the Utah Test and Training Range on August 17, 2017, during exercise Combat Archer, testing and validating the performance of ground crews, pilots, and technology while deploying precision-guided munitions. (Courtesy US Air Force.)

Six

THE 419TH FIGHTER WING

1951–PRESENT

The Air Force Reserve has maintained an active presence at Hill Air Force Base since 1947. After activation, the 419th Group controlled seven reserve units. Group pilots trained until March 1951, with preparation continuing for entry into service in July 1956, a component of Air Force flying different transport and then turbojet aircraft. In 1972, the Air Force Reserve Modernization Program upgraded the group to the Republic F-105 Thunderchief supersonic fighter-bomber. The unit was designated the 508th Fighter Group. On October 1, 1982, the group was upgraded to a wing and designated the 419th Tactical Fighter Wing. It was the last Air Force operational F-105 unit until conversion to the General Dynamics F-16 Fighting Falcons in January 1984. Over 100 of the 419th personnel were called to active duty to support Operations Desert Shield and Desert Storm during 1991–1992. Approximately 350 wing personnel deployed to Incirlik Air Base, Turkey, in 1994 to support Operation Provide Comfort II, enforcing the no-fly zone over northern Iraq. The wing returned again in 1997 to support Operation Northern Watch. In 1998, the wing's 466th Squadron deployed six F-16s to Kuwait to support Operation South Watch. In June 1999, the wing deployed back to Incirlik Air Base, becoming the first Air Force Reserve wing to use the precision strike Lightning II system. In October 2001, the wing returned to the Middle East as part of Air Expeditionary Force 8, using the Lightning II system to enforce the no-fly zone over southern Iraq. Members of the wing's 466th Fighter Squadron supported the North American Aerospace Defense Command by flying combat air patrols at undisclosed locations in the United States from December 21, 2001, to January 1, 2002, to prevent an aircraft terrorist attack. In January, the wing deployed to enforce the no-fly zone over southern Iraq. It was then activated for Operation Iraqi Freedom. In 2015, the 419th Fighter Wing became an associate unit of the 388th Fighter Wing and the first Air Force Reserve Wing to be equipped with the Lockheed Martin F-35A Lightning II.

This North American AT-6 Texan trainer was used by the 419th Fighter Wing from 1947 to 1951 to train pilots. The two-seat Texan was an advanced pilot trainer aircraft, used in preparing pilots for the transition to operational aircraft. This aircraft is on display at the Castle Air Museum. (Author's collection.)

Pictured is an AT-6 Texan on display at the Hill Aerospace Museum. The instructor sat in the rear seat, and the student pilot in the front seat. (Author's collection.)

This is a Beech AT-7 Navigator, a modification of the civilian Beech Model 18, developed in 1937 for the US Army Air Corps. It was used by the 419th Fighter Wing from 1947 to 1951 as a utility and staff transport and a bombardier trainer. This aircraft is on display at Hill Aerospace Museum. (Author's collection.)

The twin-engine aircraft shown here is a modification of the AT-7 Navigator, designated the AT-11 Kansan, flown by the wing from 1947 to 1951. The Navigator was fitted with three navigator plotting tables and one astrodome for star shots to train navigators. The aircraft is on display at the Castle Air Museum. (Author's collection.)

The Curtiss C-46 Commando seen here was flown by the 419th from 1949 to 1951. C-46s were used during the Korean War and in training for fighting in Southeast Asia. It was classified a medium-range passenger and cargo transport. The aircraft is on display at the Castle Air Museum. (Author's collection.)

Shown is a Republic F-105 Thunderchief painted in the Southeast Asia jungle camouflage scheme, flown by the wing from 1982 to 1984. It was the Air Force's heaviest supersonic fighter-bomber flown during the Vietnam War, attacking heavily defended North Vietnamese targets. The Air Force lost 397 Thunderchiefs during the war. This aircraft is on display at the Castle Air Museum. (Author's collection.)

This is an F-105 inside the Hill Aerospace Museum, shown in the Southeast Asia jungle camouflage paint scheme to blend into the terrain below the aircraft to make it more difficult for North Vietnamese pilots to locate the aircraft from above. The fighter-bomber was initially designed to drop a nuclear weapon from its internal bomb bay. It was a very effective conventional bomber in Vietnam. (Author's collection.)

Inside the maintenance hangar at Hill Air Force Base is a McDonnell Douglas F-4 Phantom fighter-bomber undergoing maintenance at the Ogden Air Logistics Center in the 1970s. (Courtesy 75th Air Base Wing historian.)

Hill Air Force Base personnel mounted this stripped F-4 airframe on a pedestal. The 419th Fighter Wing flew this type of aircraft from 1982 to 1988. (Author's collection.)

This F-4 is on display at the Hill Aerospace Museum. The supersonic fighter-bomber was capable of carrying a maximum bomb load of 16,000 pounds, and was a very effective ground attack aircraft in the Vietnam War. (Author's collection.)

The 419th Fighter Wing's first General Dynamics F-16 assigned to Hill Air Force Base is seen here on January 28, 1984. (Courtesy 75th Air Base Wing historian.)

This is a 1995 photograph of a scaled-down replica of an F-4 mounted on a display pedestal in front of the aircraft maintenance building. (Author's collection.)

Seen here are an F-16 with its canopy open and two wing auxiliary fuel tanks, with an inert air-to-air missile under each wing. The aircraft is on display at the Hill Aerospace Museum. The single-seat, multi-role, supersonic fighter was flown by the 419th Fighter Wing. (Author's collection.)

A row of F-16s is parked on the Hill Air Force Base flight line. (Courtesy 75th Air Base Wing historian.)

The 419th Fighter Wing pilots flew the Lockheed T-33 Shooting Star as a transition trainer for upgrading pilots to turbojet fighters and as a supplemental aircraft for flying hours. This aircraft is on display at the Hill Aerospace Museum. (Author's collection.)

The 419th flew the Lockheed F-80 Shooting Star, a straight-wing, post–World War II fighter. It was no match for the swept-wing Russian MiG-15, but it was a deadly air-to-ground attack fighter-bomber during the Korean war. This aircraft is on display at the Castle Air Museum. (Author's collection.)

This is a Republic F-84 Thunderjet swept-wing fighter on display at the Hill Aerospace Museum. The 419th Fighter Wing flew this type of fighter aircraft. (Author's collection.)

This Fairchild C-119 Flying Boxcar transport is on display at the Castle Air Museum. The C-119 was the standard Air Force transport in the 1950s, heavily used by the 419th during the Korean War. (Author's collection.)

This view of a C-119 Flying Boxcar shows the twin-boom tail, allowing the rear clamshell door to be opened for loading and unloading oversize cargo, or with doors removed, allowing drops of paratroops, equipment, and supplies. The aircraft is on display at Hill Aerospace Museum. (Author's collection.)

A Douglas C-124 Globemaster II transport is shown at the Hill Aerospace Museum. The C-124 was the largest mass-produced piston-powered transport. It was flown by the 419th Wing. (Author's collection.)

The Air Force only built one super heavy-lift, six pusher piston, double-deck transport: the Convair XC-99. This transport landed at Hill Air Force Base on October 17, 1950, loaded with 100,000 pounds of cargo. It flew until March 1957 before being retired. The aircraft is pictured at the Convair Plant at Strategic Air Command Carswell Air Force Base in Fort Worth, Texas. (Author's collection.)

This is a June 1990 photograph of an XC-99 transport parked in the corner of Kelly Air Force Base, Texas, with damage due to weather exposure and bird droppings. The aircraft was taken apart and shipped to the National Museum of the US Air Force for restoration, but it was determined that it would be too expensive to repair it for display. (Author's collection.)

Pictured at the Hill Aerospace Museum is a C-123 Provider, a twin-engine transport used by the 419th. The aircraft had excellent short-field landing and takeoff capabilities. (Author's collection.)

An F-35A Lightning II, the first assigned to the 419th Fighter Wing, lands at Hill Air Force Base in 2015. (Courtesy 75th Air Base Wing historian.)

Two F-35As are pictured in open-air, covered aircraft shelters, preparing to be launched for a training flight from Hill Air Force Base. (Courtesy 75th Air Base Wing historian.)

A 419th Fighter Wing F-35A is shown during a dusk takeoff from Hill Air Force Base. (Courtesy 75th Air Base Wing historian.)

This is a head-on photograph of a 419th Fighter Wing F-35A at sunset, with the mountains in the background at Hill Air Force Base. (Courtesy 419th Fighter Wing Public Affairs.)

A 419th F-35A is hooked to a tow tug and being slowly backed into a maintenance hangar bay. (Courtesy 75th Air Base Wing historian.)

This F-16 was photographed from an adjacent 419th Fighter Wing F-16. (Courtesy 75th Air Base Wing historian.)

This is a 419th Fighter Wing public affairs photograph of an F-35A with backlighting and a fog effect for a dramatic picture. (Courtesy 419th Fighter Wing Public Affairs.)

Seven

HILL AEROSPACE MUSEUM
1982–PRESENT

Hill Aerospace Museum is located on 30 acres in the northwest corner of Hill Air Force Base. The museum was founded in 1982 as part of the US Air Force Heritage Program, which the secretary of the Air Force initiated in 1970. In 1981, Hill Air Force Base began participating in the heritage program. In 1983, members of the Salt Lake City community founded the Aerospace Heritage Foundation of Utah, a private nonprofit organization, to raise funds and support the Hill Air Force Base Museum. On January 27, 1984, Maj. Gen. Marc Reynolds, the Ogden Air Logistics Center commander, hosted the initial kickoff for the proposed museum. The museum opened in 1986. By 1987, the museum and aerospace park were in operation. In 1988, through the foundation's lobbying efforts, the state of Utah provided $5 million for the construction of an administration facility and Hadley Gallery (to go with the existing Lindquist Art Gallery). Additional funds became available in 1999, and the museum's name was changed to Hill Aerospace Museum. Many of the museum's display aircraft undergo refurbishment or complete restoration, but it does not conduct remedial conservation on other collection items. This is done by the National Museum of the US Air Force Conservation Laboratory at Wright-Patterson Air Force Base in Ohio. The museum's artifact collection and aircraft are exhibited in five areas, corresponding to the museum's interpretive timeline: Beginnings, World War II, Dawn of the Jet Age, Cold War, and Keeping the Peace. The museum's missions are to research, conserve, interpret, and present the mission, history, heritage, and traditions of the US Air Force, Hill Air Force Base, and Utah aviation through engaging exhibits, educational outreach, special programs, and stewardship of the national historic collection.

This is a 2020 photograph of the entrance to Hill Aerospace Museum's indoor exhibit area, with a Douglas C-124 Globemaster II transport aircraft. (Courtesy 75th Air Base Wing historian.)

Seen here is a 1995 view of the museum and various transport aircraft in the airpark. (Author's collection.)

This is the inside of the Hill Aerospace Museum exhibit hall, with an F-16, the fighter type flown by both the 388th and 419th Fighter Wings at Hill Air Force Base. (Author's collection.)

This twin-engine Lockheed P-38 Lightning fighter is on display inside the Hill Aerospace Museum exhibit building, as a part of a World War II aircraft exhibit. (Author's collection.)

Here is a Republic F-105G Thunderchief two-seat fighter-bomber nicknamed the "Wild Weasel." During the Vietnam War, it was used to attack North Vietnamese surface-to-air missile radar sites. (Author's collection.)

Shown is a Rockwell (now Boeing) B-1B Lancer nuclear/conventional, long-range, supersonic, swing-wing bomber. (Author's collection.)

This is a Boeing B-29 Superfortress bomber parked outside the Hill Aerospace Museum building. This World War II bomber ended the war in the Pacific and fought in the Korean War against Russian turbojet fighters. (Courtesy 75th Air Base Wing historian.)

Pictured is a Boeing B-17F Flying Fortress, equipped with two .50 caliber machine guns mounted in a chin turret, designed to defend against attacking German fighters. (Author's collection.)

This is a Minuteman I ICBM enclosed transporter and erector used to move missiles from the operating missile base to a designated launch facility. It is on display in the Hill Aerospace Museum airpark. (Author's collection.)

Seen here is a Boeing Minuteman I ICBM on a display transporter at the Hill Aerospace Museum airpark. (Author's collection.)

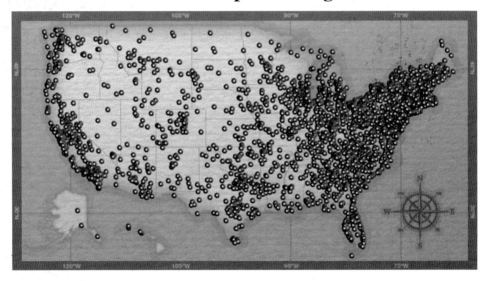